ABOVE AND BEYOND℠

THE OFFICIAL COMPANION BOOK

PRESENTED BY

Lela Nargi

Copyright © 2015 The Boeing Company. All rights reserved.

Printed in the United States of America.

First Printing December 2015

Permission to reproduce works of art, photographs, and artifacts in this volume was provided by the rights holders, when they could be identified.

Every effort was made to obtain and verify accurate identifying information for the works. Please notify The Boeing Company of any inaccuracies, and corrections will be included in future editions.

ISBN: 978-0-9968558-0-8

Published by Boeing Press

www.boeingstore.com

For more information, please e-mail customerservice@boeing.com.

Designed and Produced by Girl Friday Productions.

Designed by Zach Hooker.

Other titles available from Boeing Press:

We Own the Night: The AH-64 Apache Story

In Plane View: A Pictorial Tour of the Boeing Everett Factory

Trailblazers: The Women of The Boeing Company

The Jumbo Jet: Changing the World of Flight

Strike Fighter: The Story of the F/A-18E/F Super Hornet

Strategic Airpower: The History of Bombers

You Call, We Haul: The CH-47 Chinook Story

Remarkable: The Story of the Boeing 737

Above and Beyond is made possible by Boeing. The exhibition is produced by Evergreen Exhibitions in association with Boeing, in collaboration with NASA.

CONTENTS

INTRODUCTION

If you've picked up this book, that means you are inspired by and curious to learn more about flight and aerospace. Your curiosity is something you share with every engineer, scientist, and astronaut on the planet. It's the secret ingredient to building the future.

Everyone who worked to bring you *Above and Beyond* thought long and hard about the things that once inspired us: building airplane models, learning to work on a car engine in the garage, visiting airshows. We wanted to make sure that each part of the exhibit—"Up," "Faster," "Higher," "Farther," and "Smarter"—would resonate with you and inspire you to learn more about aerospace. The world needs curious-minded people like you to create whatever great thing comes next.

At *Above and Beyond*, visitors can fly a fighter-jet simulator, flap their wings like a bird, and ride a space elevator. The book and the exhibit give you just a little taste of the future of aerospace—and a taste of what your own future may hold if you let your inspiration and curiosity take you there.

This book, and this show, are for you!

See you in the future.

Britain's Royal Air Force Aerobatic Team performs.

Louis Blériot flies the Ambroise Goupy II.

1783   Hot Air Balloon: First flight of the Montgolfier brothers' balloon.

1794   Observation Balloon: Jean-Marie-Joseph Coutelle uses balloons for spying and intimidation during the French Revolution.

1903   Piloted Airplane: The Wright brothers fly an internal combustion–powered plane at Kitty Hawk, NC.

1909   Across the Channel: Louis Charles Joseph Blériot flying the Blériot XI is the first to fly across the English Channel in a heavier-than-air aircraft.

1909   Air Races: France holds the first international flying competition at Reims.

1912   Single-Shell Aircraft: An early racing aircraft has a single-shell, or monocoque, fuselage that reduces weight and drag.

1914   Fighter Planes: Aerial dogfighting begins in World War I.

1914–1918   Bombs Away: First use of aircraft as bombers in World War I.

1915   NACA: Congress creates the National Advisory Committee for Aeronautics, the organization from which NASA was created in 1958.

1917   Supercharged: Sanford Moss invents the turbo supercharger for pressurizing air used by engines to adjust to high altitudes.

1919   International Airmail: Bill Boeing and Eddie Hubbard fly from Seattle to Victoria, Canada.

1919   Across the Atlantic: The NC-4 seaplanes complete first transatlantic flights, followed by the flight of John Alcock and Arthur Brown.

1923   Fuel Station in the Sky: Lowell H. Smith and John P. Richter set new record for time spent in the sky, thanks to aerial refueling.

1924   Around the World: Two Douglas World Cruisers make it around the globe.

1926   Liquid Rocket Fuel: Rockets upgrade from solid to liquid fuel, which allows engines to throttle up and down and stop and start midflight.

1927   Solo Across the Atlantic: Charles Lindbergh flies solo from New York to Paris, nonstop.

1929   Women Racers: Louise Thaden wins the first all-women's air race from Santa Monica, CA, to Cleveland, OH.

1929   Flight Instruments: Jimmy Doolittle makes the first "blind" (instrument-only) flight.

1933   The all-metal, streamlined Boeing Model 247 is the world's first modern airliner.

1930s   Jet Engine: Frank Whittle and Hans von Ohain develop the jet engine independently of each other.

1930s   Radar Tracking: Leading up to World War II, Britain refines use of radar, which is used by both sides in the war.

1931   To the Stratosphere: Beginning with Swiss physicist Auguste Piccard, humans, including a woman, visit the stratosphere.

The first all-women air race, held in 1929.

An aviation worker cleans the propeller of a Mustang P-51 during World War II.

We all know about flight ace Amelia Earhart (below), but let's not forget some of the other amazing women pioneers of aviation, both before and after Earhart's solo flight across the Atlantic Ocean:

Bessie Coleman (1892–1926): She was the first African American woman to earn a private pilot's license, in 1921. But she had to go to France to get it (left).

Florence "Pancho" Barnes (1901–1975): She was a free-spirited test and stunt pilot. In 1932, she founded the Association of Motion Picture Pilots to secure fair wages for movie stunt pilots (right, bottom).

Valentina Tereshkova (1937–): In 1963, Soviet Union–born Tereshkova was the first woman to fly in space, aboard Vostok 6 (right, top).

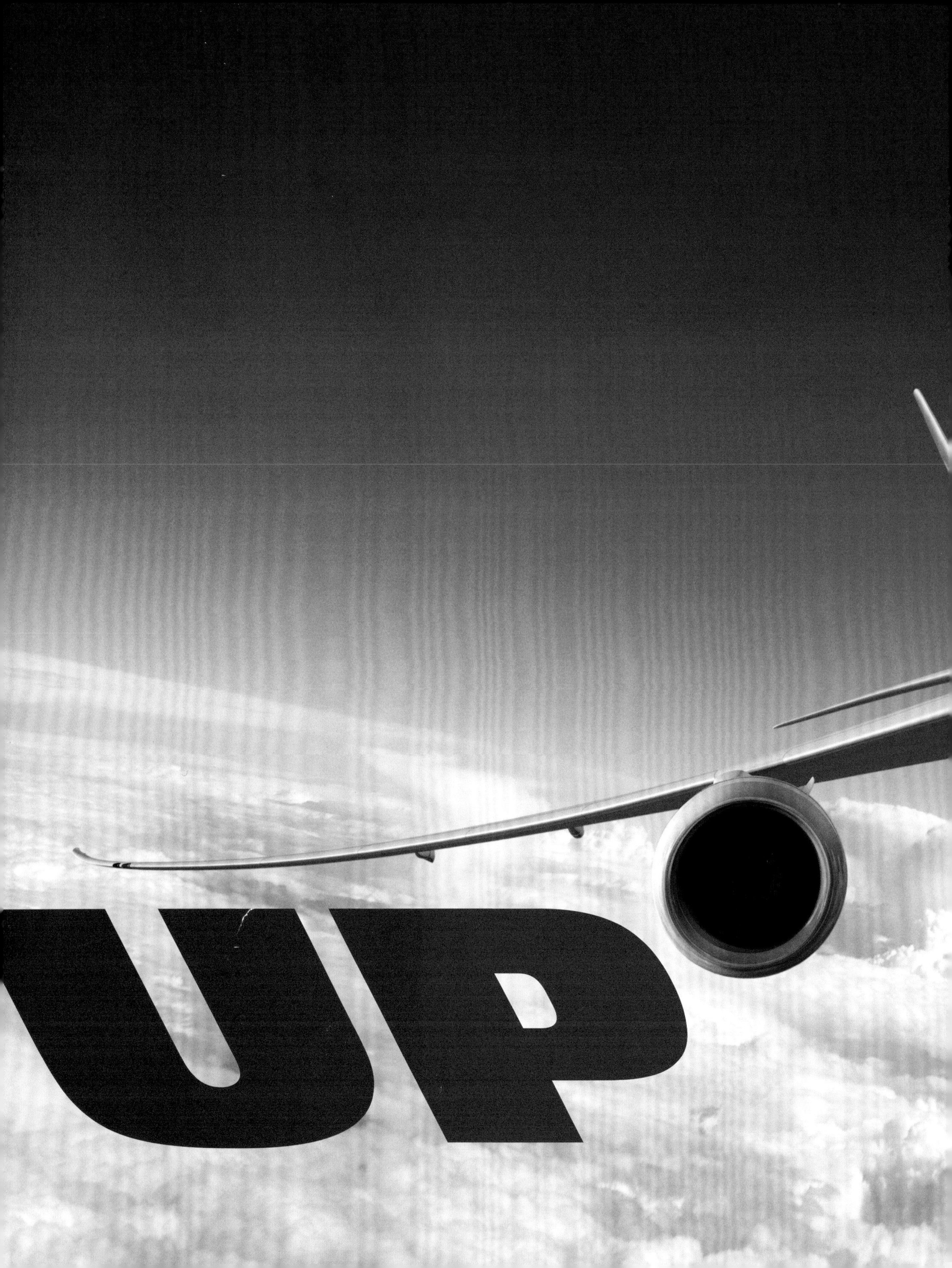
UP

GAZING UP

at the sky, our ancestors watched birds fly and imagined that, one day, they'd be able to fly, too. The very earliest designs for flying machines took their inspiration from our avian friends. But test flights of these early would-be modes of transportation ended in failure, and often injury. It takes more than flapping your feather-covered arms and jumping off a roof to get, and stay, airborne—as you can learn firsthand when you join a migrating flock of birds in "Spread Your Wings."

Whether you're a bird with specialized parts for flying—like feathers, hinged wings, and hollow bones—or a human piloting an airplane, a rocket, or a jetpack, you must contend with four forces to get yourself off the ground and into the air. These are **weight** in the form of gravity and its opposing force, **lift**, as well as **drag** and its opposing force, **thrust**. They were first identified by English engineer Sir George Cayley in 1799, and they work both for and against you when you fly.

Gravity is the force that pulls birds—and you in your aircraft—**downward**. To actually get into the air, you have to counteract this with aerodynamic lift. This is an **upward** force that pulls you away from Earth and keeps you aloft as you move above it through the air.

> ## Wow, look at you fly! It feels great to be thousands of feet off the ground, soaring through the clouds!

Wow, look at you fly! It feels great to be thousands of feet off the ground, soaring through the clouds! Not so fast. You still have to deal with drag. This is a force that causes **resistance**, slowing you down. To counteract drag, you need thrust. Thrust is a **propelling** force that moves you forward and also generates that other positive flight force, lift. Although it is possible to get into the air without thrust—engine-less hot air balloons do and so do gliders that must be launched from a high spot to get aloft—you can't maneuver around very well without it.

But how do birds—creatures that can't perform simple addition let alone complex physics equations—master the forces of flight? They're born to it. As you can experiment with in the exhibit, their wings are designed to flap up and down, which generates thrust and overcomes drag. The shape of their wings creates lift, which counteracts birds' weight and pulls them away from gravity. They also instinctively angle their bodies as they fly, to change the angle of attack on their wings. This generates thrust when they take off, as well as drag when they want to slow down. Aeronautical technology owes a lot to birds, and it continues to borrow from them all the time when designing parts for aircraft and spacecraft.

Previous page: The 787-8 Dreamliner soars. This page: Four black-headed gulls illustrate the power of flight.

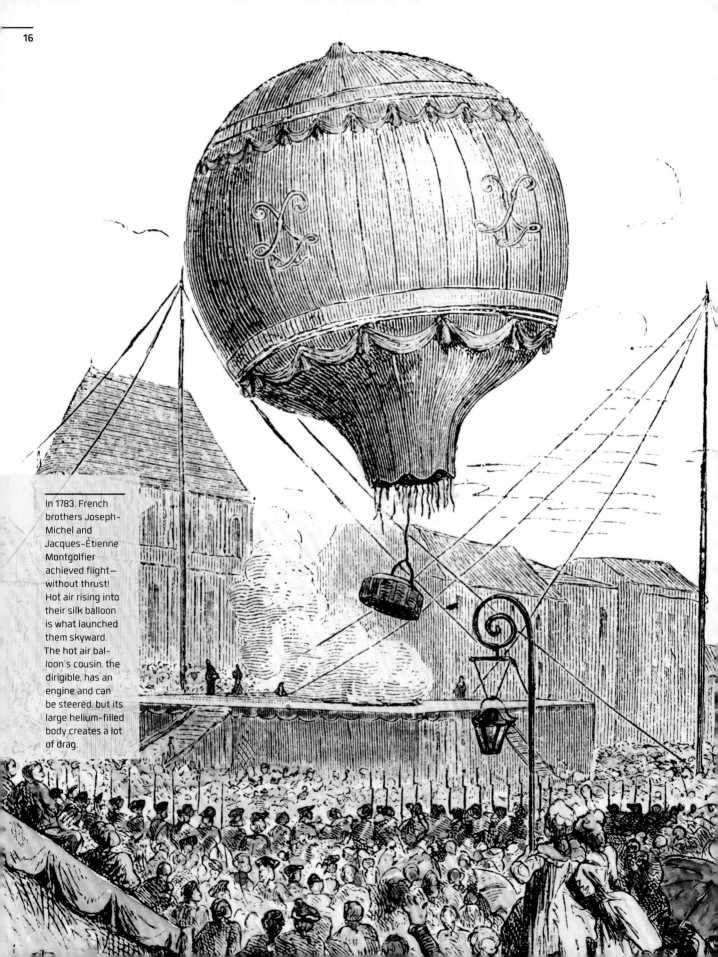

In 1783, French brothers Joseph-Michel and Jacques-Étienne Montgolfier achieved flight—without thrust! Hot air rising into their silk balloon is what launched them skyward. The hot air balloon's cousin, the dirigible, has an engine and can be steered, but its large helium-filled body creates a lot of drag.

Possibly the biggest heroes of thrust are Wilbur and Orville Wright. In 1903, their first powered, heavier-than-air machine, the Wright Flyer, launched humankind into the aerial age, making commercial passenger planes, supersonic jets, manned space vehicles, and whatever comes next, possible.

Future aircraft

Ever since we mastered the forces of flight, engineers have been devising new ways to make flying better in every way. The extremely versatile helicopter, for example, has rotors that generate lift and allow it to hover, as well as a tail rotor, which keeps it from spinning endlessly in circles. Following are some of the coolest new innovations that will make current and future aircraft faster, quieter, and more efficient.

Blended Wing Body

The 737 that flies you across the country to visit your grandmother for the holidays may get you where you're going, but its wings, which stick out from its body, produce a lot of drag. By blending the wings seamlessly into the fuselage of an airplane, such as in the case of the experimental **Boeing Phantom Works X-48** series, drag is reduced—and so are fuel use and noise.

Electric Engine

We already drive eco-friendly cars that use electricity instead of, or in addition to, fuel. Why can't we use that same technology on planes? Thanks to NASA and Boeing, someday we will! The **SUGAR Volt** ("SUGAR" is an acronym for Subsonic Ultra-Green Aircraft Research) has a hybrid engine that just may lead us to the eco-plane of the future.

VTOL

Vertical Takeoff and Landing lets airplanes with fixed wings get into the air as nimbly as a helicopter, without losing speed. The experimental **Phantom Swift** achieves VTOL with two lift fans right inside its fuselage, as well as ducted fans on the wingtips.

V Formation

In "Spread Your Wings," you can fly with migrating birds in a **V formation**. As visitors discover, the lead bird creates an upward current of air behind it, which following birds ride for free lift, conserving energy. The U.S. Air Force and Boeing have been experimenting with V formation, but with aircraft. Tests using C-17 military transport aircraft show that a V formation can cut fuel consumption by 10 to 20 percent.

C-17 Globemaster III aircraft flying in V formation.

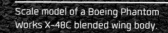

Scale model of a Boeing Phantom
Works X–48C blended wing body.

Artist rendering of a Boeing SUGAR Volt
in flight.

Snapshots From the Future

Flying cars used to be the stuff of science fiction, but as visitors are probably excited to discover in "Future Aircraft," they'll soon be a reality. The **Aero-Mobil**, by Slovakian engineer Štefan Klein, can fit in a parking space on the street and fill up with gas at the gas station. It's also extremely lightweight, thanks to its carbon-fiber shell—less than half the weight of the average compact car.

The AeroMobil will need a runway for takeoff and landing, which makes it a little impractical for everyday family use. But the **Terrafugia TF-X** four-person concept car can lift straight off the ground like a helicopter. Once in the air, its propellers will tilt forward so that it flies horizontally. Best of all: its hybrid fuel/electric engine will make it efficient.

But how will you navigate your flying car through the skies with no stop signs or traffic lights? The **myCopter** project is trying hard to figure that out.

Researchers are studying new advances in insect-inspired swarm technology, as well as autonomous flight systems. It's possible you won't have to navigate at all—your car will do all the work for you!

Want to fly solo, without any vehicle to contain you? Then the jetpack's for you. First researched by the U.S. Army in 1949, the jetpack has seen some innovation over the years:

1958: Project Grasshopper "Jump Belt"
1961: Bell Rocket Belt
1967: Bell Pogo designed for the moon
1969: Jetpack with turbojet engine
1984: Rocket packs in space and at the Olympics
2005: Rocket boots and wing suit
2009: Water-propelled jetpack
2013: Martin P-12 Jetpack with ducted fans

Clockwise from left: A 1969 test of a jet flying belt; a man in a jetpack flies over the 1984 Olympic Opening Ceremony; a Martin P-12 Jetpack in flight in 2013.

Above and below: Artist renderings of the myCopter and the Terrafugia TF-X.

Vocabulary

aerodynamics: The way air flows around an object in flight. The faster a wing flies forward, the more aerodynamic lift it creates. When lift exceeds weight, you get flight.

fuselage: The body of an airplane, usually shaped like a long cylindrical tube.

angle of attack: The angle at which a wing meets oncoming airflow. As this angle changes, the wing generates more lift or less lift. At high angles of attack, aerodynamic "stall," or loss of all lift, can be experienced.

Above and below: Aircraft like the Advanced Super Hornet are inspired by our aerial species.

Meet Dylon Rockwell, Vertical-Lift Engineer

"How helicopters and airplanes fly, they each have things called airfoils on them, except the airfoil is on the wing of an airplane, and the airfoil for a helicopter is actually on the rotor blades, spinning around very quickly to generate lift. We call that force of air being pushed down *downwash*. You can see it in movies: some guy getting saved by the Coast Guard, all this water splashing around, and

he can barely look up because the blades are chopping so fast. That's the downwash that you're seeing.

"The V-22 is a vehicle that flies like an airplane but also can convert and fly like a helicopter. It's an ugly duckling, but we save lives with these. The V-22 program revolutionized the way you can go out and save people when they've been injured in the field.

Beforehand, someone went out in an airplane to figure out where the guy was; then you had to send a helicopter to go get him. In the V-22, you're going out to find him in airplane mode, then you swoop down and pick him up in helicopter mode. Curiosity forces us to create things you've never seen before."

IN THE early 1900s, the Wright brothers' Wright Flyer achieved a top speed of some 30 mph (50 kilometers per hour). In October of 1947, U.S. Air Force test pilot Chuck Yeager flew faster than the speed of sound—which is around 700 mph—in an experimental rocket plane called the Bell X-1. Today, aerospace engineers push the limits of technology to fly at ever-faster Mach speeds.

One critical component to achieving high speeds is wing design. Every potential wing shape has an effect on how fast a craft can go, as well as how well it can fly, how much it can lift, and how easy it is to maneuver. Before wings are tested on actual planes, though, they're tested on small-scale models in wind tunnels. Exhibit visitors get a glimpse of how wind tunnels work in "Shock Waves."

In a wind tunnel, aerospace engineers force air through a tunnel at the intended speed of flight, using fans or compressors. Then using various instruments, they measure the forces produced on the model by the rushing air. They also add smoke and dyes to visualize airflow and any shock waves. Based on the data they accumulate from the tests, the engineers tweak their designs, then test them again—and again. And again.

Wind tunnels provide an amazingly accurate picture of how all kinds of full-scale craft will perform in real life. They've been indispensable in developing everything from models of a low-sonic-boom supersonic business jet to the Space Shuttle Orbiter.

Previous page: An F/A-18C Hornet breaks the sound barrier. This page: A Mach 7 wind tunnel test of the X-43A.

George Schairer, Swept Wings, and Supersonic Flight

During World War II, aeronautics engineers experimented heavily with wing design in their attempts to break the sound barrier.

Traditional **straight wings** run perpendicular to the fuselage. They provide excellent lift and stability at slow speeds. But at higher speeds, a buildup of shock waves makes the aircraft unstable. During World War II, jets outfitted with straight wings tended to fall apart at high speeds.

Boeing engineer George Schairer began experimenting with **swept wings**, which are angled to reduce drag, which in turn reduces the buildup of shock waves as planes near supersonic speeds.

Not everyone was convinced that swept wings would work. But as the war wound down, Schairer went to Germany with a group of American scientists to collect aeronautical research that had been developed by the Nazis. There, he got a lucky break: documents found in a test lab called Völkenrode held the answer to making swept wings function at supersonic speeds. Here's the letter Schairer sent to his friend and coworker, Benedict Cohn.

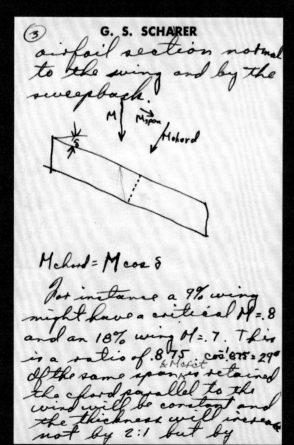

A page from George Schairer's iconic letter home.

Dear Ben,

It is hard to believe I am in Germany within a few miles of the front line. Everything is very quiet and I am living very normally in the middle of a forest. We have excellent quarters including lights, hot water, heat, electric razors, etc.

We are seeing much of German aerodynamics. They are ahead of us in a few items which I will mention. Here the Germans have been doing extensive work on high-speed aerodynamics. This has led to one very important discovery. Sweepback or sweep forward has a very large effect on critical Mach number. This is quite reasonable on second thought. The flow parallel to the wing cannot affect the critical Mach number and the component normal to the airfoil is the one of importance. Thus the critical M [Mach] is determined by the airfoil section normal to the wing and by the sweepback.

A certain amount of experimental proof exists for this sweepback effect. Only the ME163 has used it in so far as I can find out. Naturally many control and stability problems are to be encountered in using large amounts of sweep here.

I am having a fine time. I even use my electric razor wherever I go. . . . Hope things are going well for you. My best to all the gang. They are sure tops in all comparisons.

Sincerely, George

Above are the straight wings of the BT-13A Valiant, while below are the swept wings of the B-47E Stratojet. In addition to the wings, the shape and size of an aircraft's nose cone, tail, and fuselage have an effect on how well and quickly it performs.

Feeding the Need for Speed

"Full Throttle" lets exhibit visitors experiment with designing airplane components to make jets faster. Here are some innovations that made, and will make, aircraft not only speedier but also quieter, cleaner, and greener.

Delta wings are shaped like a triangle—or the Greek letter delta. They have a high sweep angle that reduces the buildup of shock waves when an aircraft reaches the speed of sound. Their big disadvantages: higher drag and poor low-speed performance (right).

Morphing wings are flexible, offering more control of an aircraft, but weight and complexity are a challenge (opposite, top).

The Concorde was the first passenger plane to fly at supersonic speed. It was decommissioned in 2003 . . . partly because its loud sonic boom precluded use over populated areas.

What's faster than **supersonic**? Hypersonic—five times faster, in fact! Until recently, consistent hypersonic speeds in aircraft were only achievable with rocket engines. Rockets need to carry their own supply of oxygen to burn along with their fuel, which makes them heavy and expensive. But a scramjet engine (below, right) can pull in oxygen from the air without having to first slow that air down to allow combustion and generate thrust, like a traditional jet engine must. One day, the Boeing **X-51A WaveRider** (opposite, bottom) could be the go-to plane for transcontinental travel.

Boeing's Phantom Eye (opposite, top) runs on clean-burning hydrogen—one tank of the stuff will power this craft for days with the only by-product, or emission, being water. Now that's seriously clean energy!

As for the sonic-boom dilemma? Below left a Boeing engineer inspects a low-boom supersonic aircraft model. One day, this just might take the place of the Concorde.

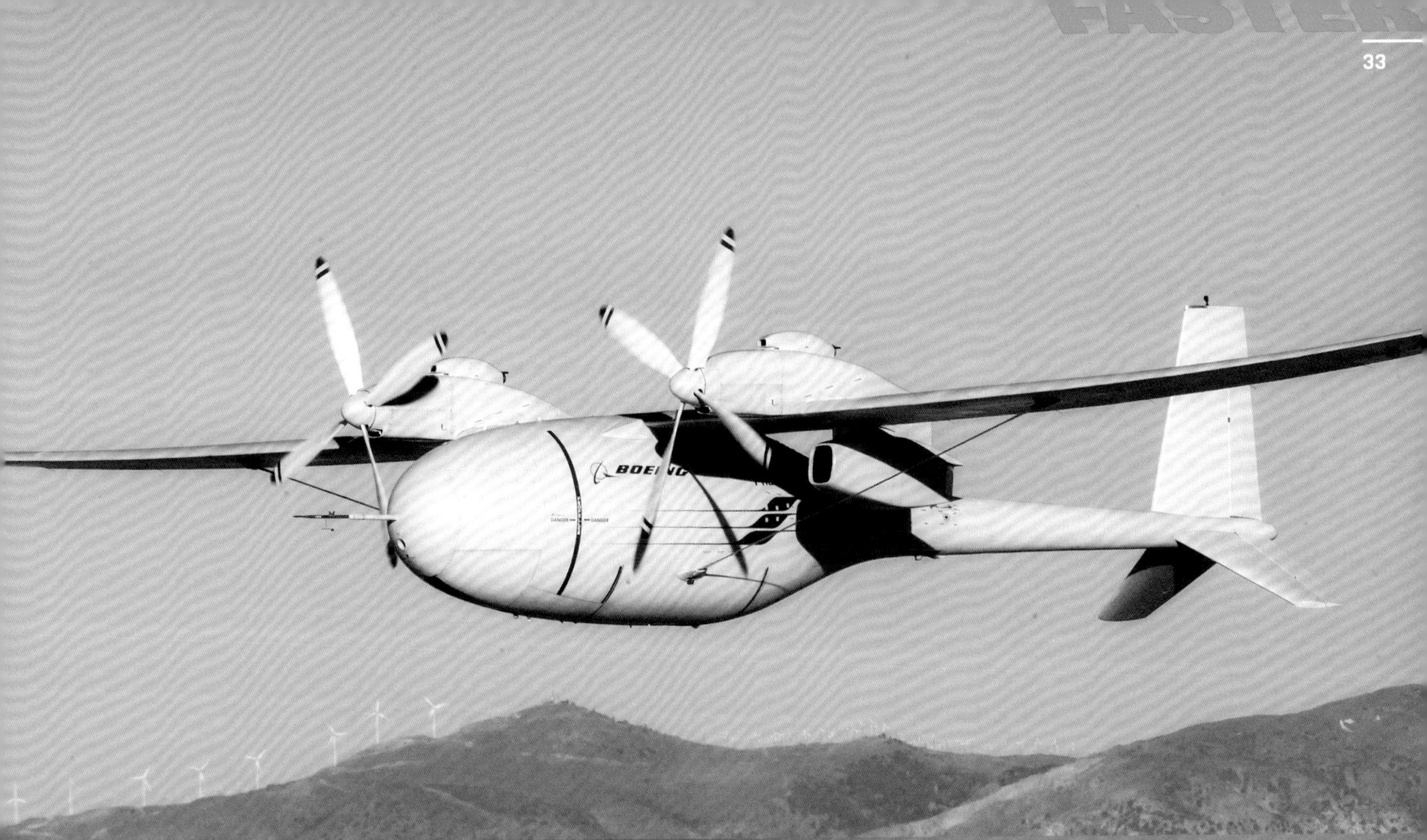

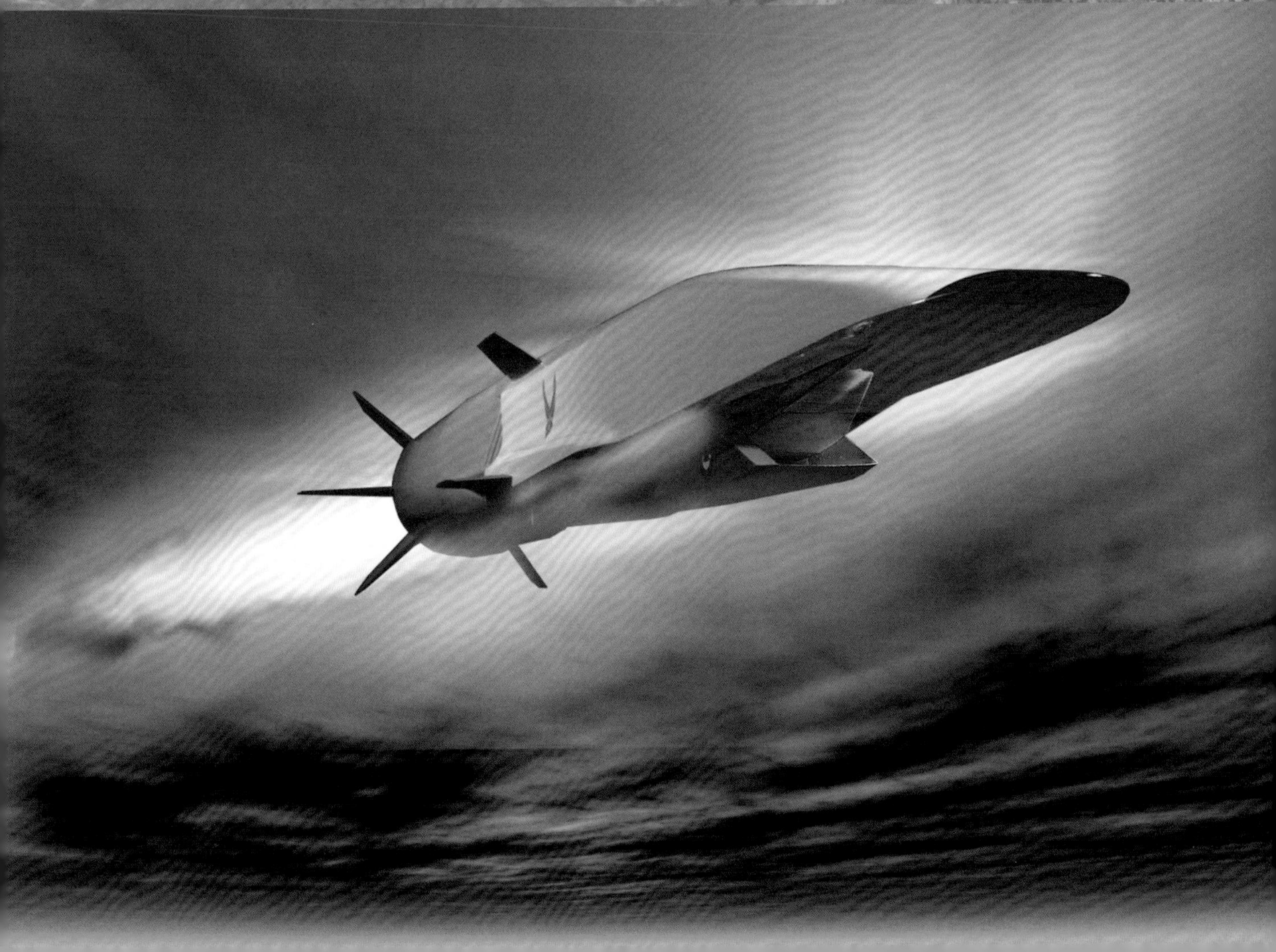

Vocabulary

Mach: The measure of the speed of sound. For example, "subsonic," which is less than the speed of sound, is < Mach 1. "Transonic," approaching the speed of sound, is Mach .75 to Mach 1. "Supersonic," above the speed of sound, is > Mach 1, Mach 2 is twice the speed of sound, and so on. "Hypersonic" is Mach 5 and higher.

shock wave: The pressure waves that build up in front of the nose and wings of an aircraft as it approaches the speed of sound. This can cause wing flutter and vibration and ultimately sonic boom.

sonic boom: The loud noise we hear on the ground when the shock waves created by an aircraft going faster than the speed of sound cause a change in pressure.

Chuck Yeager became the first man to break the sound barrier in the Bell X–1, nicknamed Glamorous Glennis.

GLAMOROUS GLENNIS

Meet Simon Bahr, Airplane Engineer

"One of the easiest ways to understand lift and the fundamental idea of how airplanes fly: You may remember when you were sitting on the passenger side of your car and rolled down your window and you stuck your hand out. And you noticed that the wind as it flowed over your hand pushed your hand up or down. You twisted your hand and the wind pushed it down, and you twisted your hand the other way and it pushed it up.

That pushing force that you feel is lift, and it's exactly what makes airplanes fly.

"Airplanes go through an incredibly long and exhausting set of tests before they are allowed to fly. One of the more exciting tests we do is the ultimate test of the wing, where we bend the wings up and have to reach one and a half times the force it will ever see without breaking anything. You start pulling it up. It's

attached on a whole bunch of pulleys and actuators, and you start pulling and pulling and eventually it just snaps.

"I love airplanes. The fact that we know how to fly, and know how to build airplanes . . . the fact that I'm sitting in this powerful piece of metal flying through the air at 500 miles an hour, there's an incredible feeling of awe."

NO SOONER had scientists and engineers and pilots mas-

tered the skies than human adventurers set their sights on an even greater challenge: propelling us out of Earth's atmosphere altogether. Writers and filmmakers had long been imagining the possibility of attaining our nearest neighbor: the moon. Jules Verne's 1865 novel *From the Earth to the Moon* features a "space cannon" that can launch three people to its surface; and in 1902, fanciful French filmmaker Georges Méliès shot his silent short *A Trip to the Moon* (opposite, top), in which a rocketful of astronomers encounter a bunch of hostile moon men. (If you read the book *The Invention of Hugo Cabret* by Brian Selznick, or saw the movie, *Hugo*, you've already been introduced to Méliès.)

In 1903, a Russian schoolteacher named Konstantin Tsiolkovsky wrote a report suggesting that space exploration by rocket was possible by using liquid propellants to achieve greater range than what was then plausible (opposite, middle). But it wasn't until the Soviet Union launched Laika the dog into orbit on the satellite Sputnik 2 in November of 1957 that touching down on the moon came to seem the stuff of possibility, not just fiction (opposite, bottom). In 1969, the United States' Apollo 11 successfully landed the very first people on the moon.

Previous page: A view of Earth from the International Space Station.

On July 16, 1969, Apollo 11 launched from Cape Kennedy carrying Neil Armstrong, Michael Collins, and Edwin "Buzz" Aldrin to the moon. Two days later, Armstrong and Aldrin became the first humans to walk on its surface.

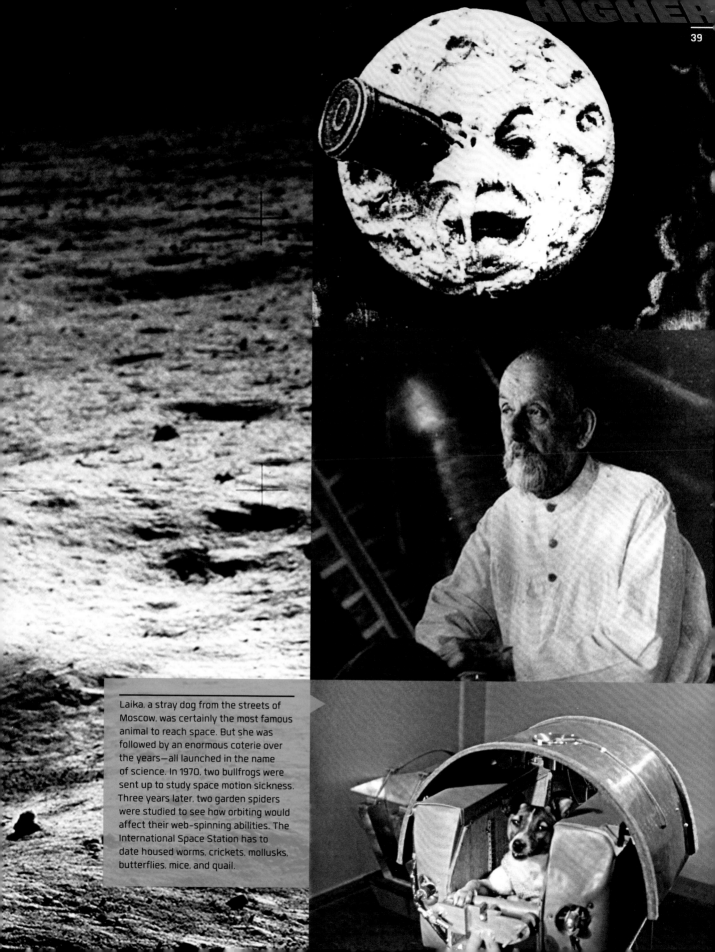

Laika, a stray dog from the streets of Moscow, was certainly the most famous animal to reach space. But she was followed by an enormous coterie over the years—all launched in the name of science. In 1970, two bullfrogs were sent up to study space motion sickness. Three years later, two garden spiders were studied to see how orbiting would affect their web-spinning abilities. The International Space Station has to date housed worms, crickets, mollusks, butterflies, mice, and quail.

Almost 50 years later, scientists and researchers from 18 countries have managed to continuously occupy the International Space Station, a micro-gravity laboratory 220 miles (350 kilometers) above Earth that you hear a lot about in *Above and Beyond*. NASA is busily working out the details for sending astronauts to Mars. A number of companies are currently working to fly folks to the moon and back—if they've got big bucks to pay for the experience. And fueling the imaginations of space scientists the world over is the space elevator.

Envisioned as a giant cable that will be tethered to the Earth's equator and extend 62,000 miles into orbit, the space elevator will transport crews en route to deep-space missions past the Kármán line and the farthest of the communications satellites, to a maximum temperature of 1,350 degrees Fahrenheit (730 degrees Celsius)—all without having to send costly rockets from Earth.

Elevator travelers leave from an anchor station somewhere in the Pacific Ocean, then ascend a cable called a "climber" all the way to the end, where a counterweight keeps the whole mechanism taut and steady as the rotation of Earth drags it along. You can take a virtual trip in "Elevator to Space." Are you ready to try it in reality?

A Japanese construction company believes it has the capability to get a space elevator up and running by the year 2050. Why will it take so long? Materials. So far, there's no substance strong enough to make the cable. Scientists have high hopes for carbon nanotubes, which have been manufactured in a lab to be 13 times stronger than steel.

The International Space Station, photographed by a crew member after departing.

Tower to Space Timeline

A space elevator may still be decades away from fruition. But people have been dreaming of reaching into the heavens for decades, beginning with Konstantin Tsiolkovsky, who was inspired to the concept in 1895 after a visit to the Eiffel Tower in Paris (below). His driving idea was to build a very tall tower, and many space dreamers after him built on that theme. Here's how the concept evolved over time:

1960: A Russian engineer named Yuri Artsutanov was the first person to propose a structure formed from a tether—to make an elevator rather than a tower.

1975: The first real scientific study of what it would take to get a space elevator functional was published by Jerome Pearson, an engineer for the U.S. Air Force Flight Dynamics Laboratory.

1991: Carbon nanotubes (below) were discovered by a Japanese physicist named Sumio Iijima. The future of space elevators started to look bright!

2014: Japanese construction firm Obayashi claimed it would be able to build a long enough cable for a space elevator out of carbon nanotubes by 2030 and get the elevator itself up and running by 2050. Are you ready to soar?

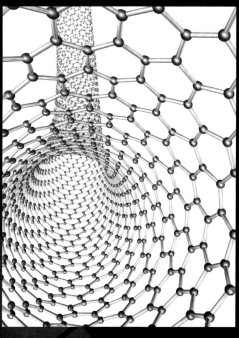

Vocabulary

microgravity: A condition in which gravity is very low and people and objects behave as though weightless. You can see this in photos and videos of astronauts floating in space.

Kármán line: At 62 miles (100 kilometers) above Earth, the invisible boundary between our atmosphere and space.

carbon nanotube: A miniature hollow cylinder that is grown in a laboratory and incredibly strong.

Meet Tony Castilleja, Manned Spaceflight Engineer

"The Boeing CST-100 Starliner is a cruise-based transportation system. It really is going to revolutionize the way that we do space travel. It's completely autonomous. The human is there just in case, as a fail-safe system to take over if something goes wrong. We can actually launch, dock, and come home completely autonomously, without any interaction of the human and the vehicle. We're bringing a completely different world of innovation that's housed in the batteries, in the flight software, in the displays, in the interaction of the human with the vehicle.

"In the Boeing CST-100 program, I work with engineers who worked on Apollo, who worked on the International Space Station, who worked on the space shuttle. When we look at building this next generation of space vehicle, we also have to look back at the data. The data trumps all. And when we look back, the optimized shape is a capsule design. The Apollo capsule, the Gemini capsule, that design was

key. You're going from 0 miles an hour on the ground in Florida, to nearly 18,000 miles an hour. So the most important concept is aerodynamics. How does this vehicle and its shape and its design interact with the air around it? Then you get to orbit. Now you're in the environment of space. Thermal interactions take over because every 90 minutes, you're revolving around Earth. Every 90 minutes you go from −250 degrees to +250 degrees Fahrenheit. It's a very tough environment. It's like going from Death Valley to the North Pole every 90 minutes. So how can a structure like this vehicle itself maintain its integrity throughout the entire flight?

"When we look at this space capsule, you also have to think about every single engineer who's an expert in each different field, from a window, to a thermal protection system, to an abort thruster, to air around it. Everyone has to work together to be able to create what you see."

An artist rendering of the Crew Space Transportation CST-100 Starliner.

IT'S THE closest planet to Earth. But at a distance of 34 million miles, or 54,717,696 kilometers (when our planetary orbits are lined up most favorably, that is), Mars, for decades, seemed beyond our grasp—unless we were reading science fiction. But as exhibit visitors learn in "Marathon to Mars," all that is about to change. And it will change in your lifetime!

So much about what we used to believe about the Red Planet was wrong: scientists once speculated that it was incredibly hot, and completely barren. But thanks to 40 years of study by robotic explorers like *Spirit* and *Curiosity*, which have sent back images and readings, we now know that the average annual temperature on Mars is frigidly cold (−81 degrees Fahrenheit, or −62.78 degrees Celsius) and that liquid water flows on its surface. And most incredibly of all, in the 2030s, humans should have the chance to experience all of this—and so much more that's still unknown about the planet—for themselves.

NASA's manned mission to Mars will be the most complicated space voyage that's ever been attempted. Absolutely nothing can be left to chance, starting with when to launch and how long to stay. A short visit of 30 days would be the least taxing on astronauts' bodies. But because Earth and Mars will have orbited away from each other in that amount of time, the flight back home would take 430 days, for a total trip duration of 640 days. If astronauts stayed put on the planet for 550 days, until the planets were lined up just right again, then the return trip to Earth would be 180 days, for a total trip duration of 910 days—that's about two and a half years!

While NASA scientists contemplate the pros and cons of each trip length, they also have to contend with other concerns. One of the greatest of these has to do with the mental well-being of the astronauts. How do you keep them from being homesick and bored? How many games can they play in all the time it takes them to journey there? How much "busywork" can they possibly tolerate? One way to avoid the issue altogether is to put them to sleep! Temporarily, that is, in a state of suspended animation in which their bodies will be cooled to induce sleep, and all nutrition will be pumped in, and all waste pumped out.

Another challenge is how to minimize the cell damage from the radiation astronauts will be subjected to as they travel, coming both from the sun and outside our solar system. A high-antioxidant diet of foods like blueberries, kale, and dark chocolate should help.

Finally, how will the astronauts' bodies fair in microgravity conditions for that extensive amount of time? In conditions of low gravity, bones and muscles weaken, even with exercise. The solution just might be creating a spacecraft that will spin to mimic Earth's gravity! (The Hermes spacecraft in *The Martian* has a section that spins like this.)

Previous page: The space launch system rises above the clouds. This page: an image of the planet Mars.

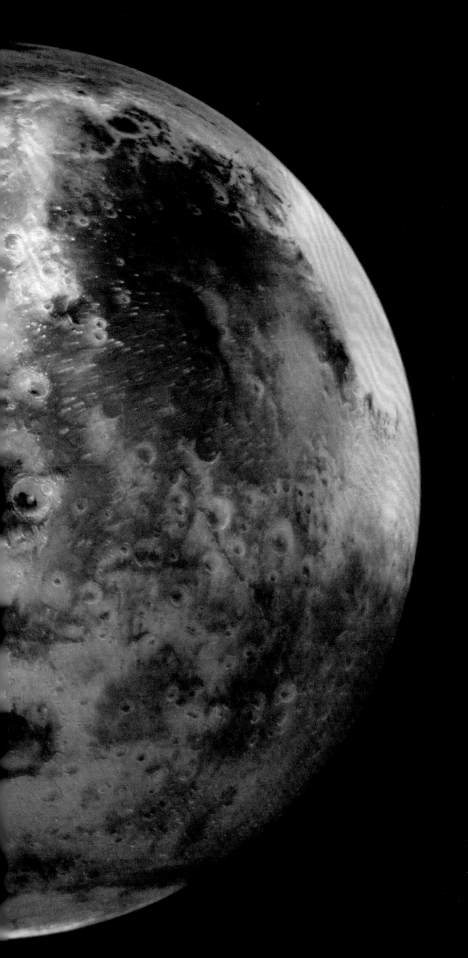

New Duds for Mars

You know all those photos you've seen of astronauts wearing bulky, hard-to-walk-in space suits, helmets, and boots? For the Mars mission, these might be totally obsolete! This isn't because conditions on Mars are any more favorable to humans than the conditions on the moon or outside the International Space Station—there's still not enough oxygen to breathe, and not enough air pressure, either.

But moving around in those space suits is extremely hard work, and incredibly awkward, especially when astronauts have to handle tools while they're working on repairs (you can simulate what it's like with the Activity on page 82).

A professor at the Massachusetts Institute of Technology named Dava Newman has invented a sleek, stretchy space suit called the BioSuit™ that could offer astronauts on the Mars mission a lot more mobility—and be a lot less frustrating when they have to do things like tend crops and take photographs of their surroundings (right).

The trip to and from the Red Planet also warrants its own special "Skinsuit," created to counteract the effects of low gravity on leg bones and spines. The Skinsuit is a compression suit that basically squeezes an astronaut's body to health (far right)!

A northward view of the Wdowiak Ridge on Mars.

How will we get to Mars? As visitors witness in "Future Spacecraft," it will take the help of lots and lots of highly specialized parts.

Space Taxi: Astronauts may one day launch to Mars from the International Space Station (ISS). The Space Taxi will get them there! Boeing's Crew Space Transportation-100 Starliner is designed to fly over and over to the ISS and other low Earth-orbit destinations—on autopilot (right, top).

Space Launch System (SLS): The most powerful rocket ever designed is just what's needed to launch humans and complex robotic missions outside Earth's orbit. On a mission to Mars, its four engines and two rocket boosters will generate 9.2 million pounds (almost 41 million Newtons) of thrust on liftoff. At the top of the SLS is the Orion Capsule (right, middle).

Orion Capsule: Up to four astronauts will sit inside this capsule built by Lockheed Martin, both to and from Mars. It's equipped with an advanced heat shield to keep astronauts safe and the fastest computers ever designed for a human-inhabited space vehicle, and it attaches to a deep-space habitat that contains a kitchen and sleeping quarters (right, bottom).

Robotic Lander: Astronauts will need a special vehicle just for getting them to the surface of Mars. It might very well look something like the Morpheus Lander (opposite, left).

Carbon Fiber: All of the elements that go to Mars will have to be lightweight and extremely durable. Carbon fiber might be the perfect material. Made from filaments that are thinner than human hair, it's formed into tape or sheets, molded into shape and "cured" or cooked in an autoclave, or high-temperature, high-pressure oven. Another lightweight possibility is microlattice. It's made of hollow polymer tubes a thousand times thinner than human hair (opposite, top right)!

Microwave Beams: Forget using fuel to get all that technology into space. An engineer named Leik Myrabo thinks we can use laser beams to thrust a vehicle called a Lightcraft out of the atmosphere (opposite, bottom right)!

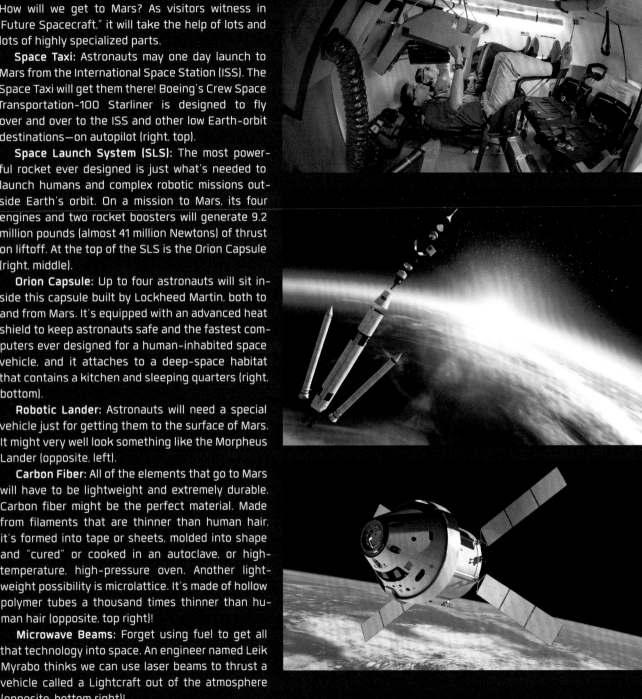

Clockwise from top: The view of Mars from *Curiosity* rover; a *Curiosity* self-portrait; an artist rendering of the Mars rover, *Spirit*; a close-up of *Curiosity*.

CURIOSITY

Vocabulary

radiation: A powerful and dangerous kind of energy that, in space, originates from the sun and other sources.

filament: A slender, threadlike fiber.

Inside the barrel of the SLS.

Meet Myron Fletcher, Rocket Propulsion Engineer

"I'm a rocket propulsion engineer working on the Space Launch System program. The SLS is a rocket that has the potential to take humans into deep-space exploration, which has never been done before.

"The biggest mission that we're shooting for is Mars. It's going to take a lot [to get there]. The first thing you're going to need is a big rocket: the SLS. You're going to need the capsule to get you in and out of Earth's atmosphere: Orion. You're going to need some type of propulsion system to get you to Mars, you're going to need a Mars lander to land on Mars, you're going to need a vehicle to get off Mars, then you're going to need a way to get back home.

"[Earth's] gravity is a problem. The biggest thing is when you jump, you come back down, so to counter that you need something more powerful than that jump. The simple principle of rockets is F = ma, which is force equals mass times acceleration. So you have a mass, and you have an acceleration, and you have this object that's being pushed, so it takes a lot of thrust to get something that big off the ground. SLS is about 321 feet [98 meters]. That's taller than the Statue of Liberty, so we're talking about a massive vehicle able to carry a massive amount of payload, which is what sets SLS apart from any other program.

"Growing up, I heard the sky was the limit—and SLS totally disrupts that. We may be shooting for Mars, but we may find more. We get to explore something that's never been explored before and the opportunities are just limitless."

THE WORLD

is filled with things that humans once could only dream of. Airplanes carry us across oceans, rockets take us to space, people can travel five times faster than the speed of sound. As soon as we get used to one new impossible possibility, another one emerges to take its place: a trip to Mars, an elevator to the sky, a spacecraft fueled by laser beams. And there is so much more in store for us, because innovation never slows down. It feeds on itself, creating endless variety. And every day, the innovators themselves get assistance from technology that is ever smarter.

Aerospace engineers working today have all kinds of futuristic-sounding tools to help them design and build: virtual reality rooms that let them collaborate with other engineers around the world; three-dimensional, or 3-D, printers that turn their designs into almost-instant prototypes; robots that do a lot of the most dangerous work for them.

Even the "simple" airplane is undergoing significant smart upgrades. Many of them emanate from space, brought to us by satellites! Satellites can keep your plane on schedule, show your pilot the most efficient route to fly to the airport, and alert her to any upcoming weather problems. What does all that mean for you? A lot less time spent traveling, and a lot more time on the ground having fun.

Previous page: An ESA Earth Explorer in orbit. This page: A satellite view of farmland in Bolivia.

Curiosity printed from a 3-D printer.

Here are some of the coolest new satellite concepts that are currently in the works:

Dove Satellite: Imagined as a grouping of hundreds of small satellites flocked around Earth in order to photograph it, documenting planetary changes (far right).

EDSN Satellite: Also imagined as a grouping, this is one of about eight "smallsats" that can be controlled via smartphone (right).

Sprite Satellite: A satellite designed to launch 100 tiny satellites into space (middle).

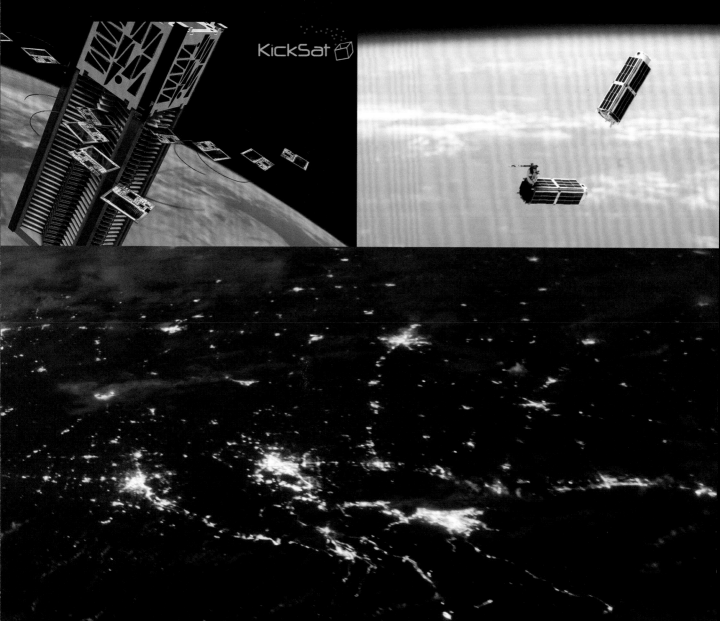

Clean Up Your Junk

It's hard to imagine that there could be any downside to all of humankind's adventures into space. But there is at least one, and it's a doozy.

Orbital debris, also known as space junk, is anything that's been left behind in Earth's orbit that can't maneuver on its own. There's a whole lot of it out there—millions of objects, in fact, including used-up pieces of rockets and satellites that ran out of fuel before they could make it home.

At *Above and Beyond*, visitors pretend to be space junkers, cleaning up small bits such as paint chips, medium-size bits such as tools, and large bits such as those rocket bodies from the atmosphere. In reality, the best defense against space junk—which can collide with still-working satellites and functional craft like the International Space Station and cause significant damage—is to track it (though this only works for pieces larger than 4 inches, or 10.16 centimeters, in diameter), avoid it if possible, and deflect it.

So far, the best deflection has been provided by the Whipple Shield, which is not unlike the shield used on the ISS. Made of multiple layers of thin aluminum, its outermost layer can break a fragment of space junk moving at 22,000 mph (35,400 kilometers per hour) into even smaller fragments before being stopped by the inner layer.

Right and far right: Tests on the impact of orbital debris on spacecraft shields.

A rendering of orbital debris based on actual data.

Roboflyers Are Coming!

In many places, roboflyers have already arrived. They take photographs at sporting events. They track whale migrations [ScanEagle]. And some of them even help fight fires.

But the future of unmanned aerial vehicles (UAVs) is gearing up to be a lot bigger and better! And in some instances, it will rely on biomimicry to guide it—that's when engineers and biologists get together to explore problems whose solutions might be found in nature. The answers to difficult technological questions can often be found outside in the world around us, perfected by millions of years of trial and error.

For example, some day in the future, RoboBees designed at Harvard University may take the place of real bees to pollinate crops for astronauts on Mars, using flapping wings to perform delicate maneuvers and swarm intelligence to communicate with one another.

Hummingbirds are small, lightweight, and very maneuverable. They can change directions quickly, hover, and fly sideways. They inspired the design of the Nano Hummingbird, a 6-inch remote-controlled surveillance UAV that has flapping wings, can change directions quickly, and navigate tight spaces. This craft is so lightweight that even when carrying a small camera, it weighs less than an AA battery.

A fixed-wing UAV, like AeroVironment's Raven, glides birdlike through the air carrying thermal-imaging cameras to count bird populations for the U.S. Geological Survey.

Miniature flying robots, as small as a penny.

Vocabulary

smallsat: Simply, a small satellite, which performs short missions and then burns up in Earth's atmosphere when its work is complete. Look, Ma, no space junk!

unmanned: Un-crewed—without humans on board.

biomimicry: The design of materials and structures that use things from nature as inspiration.

swarm intelligence: The behavior of a group—like bees—rather than of one entity in that group.

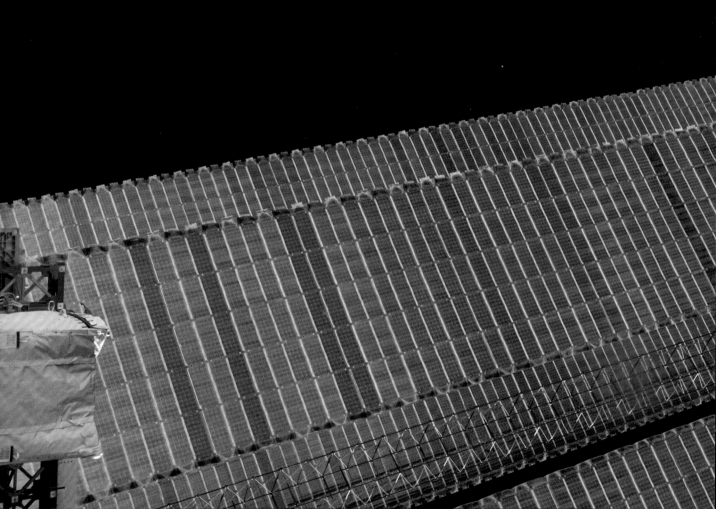

Meet Tricia Hevers, Satellite Engineer

"A satellite is a man-made object that orbits Earth or other planets. There are satellites that do everything from deliver television to people or make sure your cell phone is connected, and there are other satellites that actually monitor the weather, so when you go to see what the weather's going to be like tomorrow, satellites are the ones that bring that information to the ground.

"This is a really exciting time to be interested in spacecraft. On my program, we are building everything from satellites as small as loaves of bread, called nano-satellites, up to about 1,000 kilograms (2,205 pounds), which is the size of the refrigerator that you have in your kitchen. One of the cool things that I've learned since starting work is that it takes a lot of people to launch a satellite into space; you need people who are good at talking to other people, you need people who are good writers, you need people, who, sure, are good at math and science—but you also need people who are imaginative, to push those boundaries of what are the next things we're going to do in space.

"I do everything from the very math-heavy writing of control algorithms, and making sure the satellite is pointing in the right place, to making sure that different people in my program are talking to one another. Our vision for small satellites is that you can do a wide variety of missions—everything from creating the next GPS constellation, using small spacecraft to monitor weather, and even doing satellite-to-satellite repair services. Almost anything you can imagine is fair game in the small satellite market. It gives everyday people more access to space."

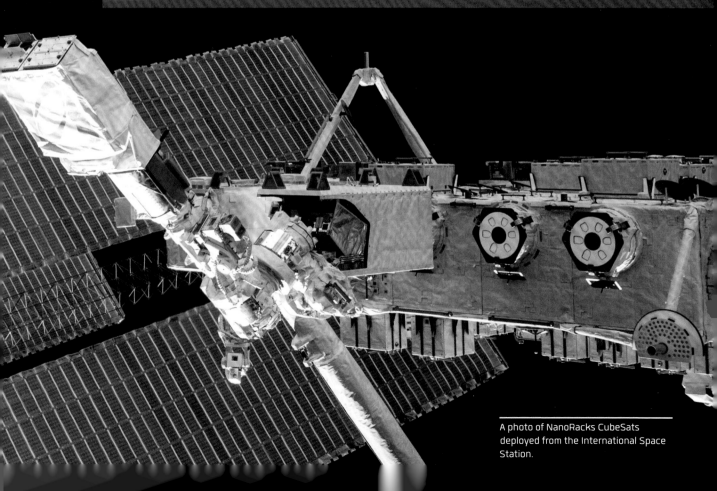

A photo of NanoRacks CubeSats deployed from the International Space Station.

S ALOFT

Dreaming Big

Where do you dream of traveling to? Is it Mars? Or the exoplanet Kapteyn b? Or one of the 60 moons of Jupiter? Would you like to test the latest in hypersonic technology? Or build the next generation of deep-space explorers?

The Boeing engineers who work on projects as diverse as passenger and jet planes to satellites and Mars missions started out with big dreams like yours. What inspired them as kids to become engineers—and what do they think the future of aeronautics has in store for us? Read on to find out.

The Past

Joseph, engineer: I remember when I was four or five years old, taking my first international trip across the Pacific to Asia. I flew on a 747 and I thought, man, I was this little guy sitting on this big jumbo jet. It was quite eye-opening.

Jacklyn, engineer: If I wasn't an engineer, I'd probably be an artist. I love to draw, and I get to use that all the time in my work.

Jillian, aerospace engineer: As a kid, I wanted to be an astronaut. I went to Space Camp, did the whole thing. When I got to college, there's no curriculum to become an astronaut, so I chose to be an aerospace engineer, which is the closest you can come without actually going to space.

Troy, propulsion design engineer: I am an unequivocal sci-fi nerd. I remember *Robotech* when I was little—I watched that. I watched *Star Wars* and *Star Trek*. I want to make that real.

Anastasia, engineer: When I was growing up, I loved to collect little things, gadgets and widgets. Once I had all the pieces that I felt like I needed, I would start to create this machine. Going through that process, watching everything work together, and always looking for that next piece that I could add to make it move a little farther or have an extra step—I loved doing that.

Adam, airplane research and development: I know I had an astronaut helmet that I made out of a cardboard box that I wore up to an age that you certainly shouldn't have been wearing a cardboard box on your head.

Liz, experimental systems engineer: I remember we had K'Nex. We had wheels and a little motor, and I used to build cars and try and send them down the hall to my sister's room, and put notes and whatever on it, and she'd send it back, maybe. If she was feeling nice, she'd send it back.

The Future

Jacklyn: In 100 years, we will fly to distant planets or comets to get new raw materials that we can use in our airplanes.

Jillian: In 100 years, we'll fly to Mars to live there.

Farah, vehicle engineer: I think maybe we can go on vacation to some new planet. Maybe we have to go with oxygen and it will be uncomfortable. But it will be awesome.

Madeline, intelligent device engineer: I think in 100 years we'll fly to a black hole to discover what's on the other side.

Anastasia: We'll travel to the moon to visit our families.

Adam: In 100 years we'll fly to Italy for lunch.

Troy: You know what? I think I'm going to go to Lagrange Point No. 3, on the other side of the moon. Just go there for a weekend, hang out, see some cool fireworks in space. And then just come back.

Have you been inspired by *Above and Beyond* and all the things that engineers can do? Visit www.boeing.com/our-future to find activities and watch videos, and to learn even more about the history and the future of flight.

Get Inspired!

These three design challenges from Curiosity Machine take their inspiration from the real-life challenges faced every day by engineers at Boeing. They were all built by kids and their families. What are you inspired to build after reading *Above and Beyond* and visiting the exhibit?

Air-Powered Spinning Machine. You need thrust for both airplane and rocket flight. You can also use thrust to steer your craft.

CURIOSITY
MACHINE

Lightweight Airplane Wing. An airplane needs to be light, so it uses as little fuel as possible, and wings are an important component of a plane's design.

Airplane Powered by Stored Energy. Traditional airplane and rocket engines need a lot of chemical fuel in order to move. But by using stored energy, an engine can be a lot more eco-friendly.

ACTIVITY: LIGHTWEIGHT AIRPLANE WING

In this activity, you'll build a lightweight airplane wing inspired by what you learned in *Above and Beyond*. Find out more about this project, and watch the video, at www.curiositymachine.org.

What you'll need:

Different types of paper
Tape
Straws
Sticks
String
Paper clips
Paper cup
Marbles
Small piece of wood
C-clamp

Start by thinking about the different wing shapes you saw in *Above and Beyond*. What will your wing look like? Sketch it out.

1. Pick a few lightweight materials you want to use to form your plane wing. Remember that you want your wing to be as light as possible.

2. Build the outside body of your airplane wing. Remember that the outside body of the wing needs to hold the inner support structure. Make sure it is at least 12 inches long.

3. Build the inner support structure of the airplane wing. Think of how you can use soft materials like paper or straws to make it stronger. Hint: You can fold and shape the paper so that it can support more weight.

4. You can take advantage of geometric shapes like triangles or diamond shapes to strengthen your wing. You can make these shapes by cutting note cards into stripes and making small slits that can be used to connect the pieces together.

5. Add additional support structures to the inside of the airplane wing. Insert the support structures into the outside body of the wing if needed.

6. Connect the wing to a table using a small piece of wood and a clamp. (If you don't have these materials at hand, think of another way you can connect the wing to the table, or find a friend who will hold your wing for testing).

7. Attach the paper cup to the wing about two inches from the end of the wing (away from the table).

8. Load marbles into the cup to test your design.

CURIOSITY
MACHINE

ACTIVITY: BIOMIMICRY

Based on what you've learned about biomimicry, design and draw a craft of the future based on a biological creature. That creature could be an insect, like the bees that inspired the RoboBee. It could be a bird, like the hummingbird that formed the basis of the Nano Hummingbird. Or it could be some other animal entirely—or a plant, or water! What will you choose for inspiration and why? What special abilities will your craft have? What will you name it?

All you need to complete this activity is a piece of paper, a pencil, and your own imagination.

ACTIVITY: MAKE A MARS SUIT

If humans of the future want to go out for a walk on Mars, they will need special outerwear to protect them. In this activity, you'll put together a simulated space suit, and then figure out how to improve it.

What you'll need:

1 pair of large, elastic-waist pants, like sweatpants
2 sections of 4" diameter dryer vent hose (commonly found at hardware stores)
2 large pairs of thick socks, in addition to the socks you're already wearing
1 pair of large galoshes or boots
2 large sweaters
1 large puffy winter coat
2 pairs of gloves:
    1 extra-large pair of garden or work gloves, and 1 smaller pair to fit inside the larger pair
1 helmet
1 pair of sunglasses or safety goggles
Disinfecting wipes
Duct tape
Scissors
Stopwatch
8 closed safety pins in a Ziploc Baggie
Nuts and bolts, 3 different sizes, separated

Begin fully dressed in your own clothes but remove your shoes. If you have a friend with you, he or she can be your Equipment Engineer. Now:

1. Put on the sweatpants over your pants.
2. Slide a piece of hose over each leg and up past the knee. Secure the hose to the pants with tape.
3. Step into the boots.
4. Put on the sweaters over your own shirt.
5. Put on the winter coat and zip/button it closed.
6. Put on the sunglasses or goggles.
7. Put the helmet on.
8. Put on the two pairs of gloves, beginning with the smaller pair.

Here are your tasks:

1. Beginning in a standing position, complete 15 jumping jacks, then 15 push-ups. Return to standing.
2. Crawl across the floor. Begin and end in a standing position.
3. Unzip the Baggie of safety pins, open each pin, link the pins together like a bracelet, return them to the Baggie, and then seal it.
4. Write down the following information that's been sent to you from NASA, as dictated by your Equipment Engineer: "The temperature on Mars may range from a high of about 70 degrees Fahrenheit at noon at the equator in the summer, or a low of about -225 degrees Fahrenheit at the poles."
5. Begin with the nuts and bolts separated and lying on the floor. Pick up all of the pieces and connect the nuts to their bolts.

Which of the tasks was the most uncomfortable or difficult to complete? Why? What could have helped?

Use this information to design a prototype suit to be worn by the first Mars explorers. First, sketch the front and back view of your design. Label the diagram to show where each of its features will be and where your equipment will be stored. Then, write out an explanation for each feature on your suit, describing how it will work. Get ready to suit up to try out your new design!

Acknowledgments

Big, bountiful thanks to Leslie Miller and Kristin Mehus-Roe at Girl Friday Productions for thinking I'd be right for this project—even working at breakneck pace. And to everyone who jumped in to help with all the right answers, especially Jim Newcomb, Jenna McMullin, and Deepa Gupta at Boeing; Susan Kirch at Right Brainiacs; and Amy Kim at Iridescent.

Author Bio

Lela Nargi is an author and journalist who writes a lot about science for kids. Having grown up listening to the stories of her airplane engineer father, working on this book felt a little like coming home. She lives in Brooklyn with her husband, her daughter, an old dog, and a young rabbit.

Image Credits

All images courtesy of and copyright © The Boeing Company unless noted below.

Engineer profiles are featured in the Aerospace Engineering Collection on PBS LearningMedia™ (www.pbslearningmedia.org/collection/aeroeng/) ©2015 WGBH Educational Foundation and The Documentary Group. All Rights Reserved.

Page 1: © NASA, ESA, M. West (ESO, Chile), and CXC/Penn State University/G. Garmire, et al.
Pages 2–3: © NASA/ESA, SSC/CXC/STScI

Introduction
Pages 4–5: © Ed Norton/Lonely Planet Images/Getty Images
Pages 6–7: © Hulton Archive/Stringer/Getty Images
Pages 8–9: Courtesy Library of Congress
Page 8: © AISA/Everett Collection; © Lordprice Collection/Alamy Stock Photo; © Hulton Archive/Stringer/Getty Images; © Keystone-France/Gamma-Keystone via Getty Images; © PhotoQuest/Getty Images
Page 9: Courtesy NASA; Courtesy NASA; © Keystone-France/Gamma-Keystone via Getty Images
Page 10: © Keystone-France/Gamma-Keystone via Getty Images
Pages 10–11: © SPUTNIK/Alamy Stock Photo; Courtesy Michael Ochs Archives/Getty Images
Page 11: © PF-(aircraft)/Alamy Stock Photo; Courtesy San Diego Air & Space Museum

Chapter 1: Up
Pages 14–15: © Russ Rohde/Getty Images
Page 16: © North Wind Picture Archives/Alamy Stock Photo
Page 17: Courtesy Library of Congress
Page 18: © USAF/Alamy
Page 20: © AP Photo/Bob Daughtry; © AP Photo/Rusty Kennedy; Courtesy Martin Jetpacks

Page 21: Courtesy Gareth Padfield, Flight Stability and Control/myCopter — Enabling Technologies for Personal Aerial Transportation Systems; Courtesy Terrafugia/www.terrafugia.com
Page 22: © Cappi Thompson/Getty Images

Chapter 2: Faster
Pages 24–25: © US Navy Photo/Alamy
Pages 26–27: © Jeff Caplan/NASA via Getty Images
Page 28: © The George Schairer Engineering Collection/The Museum of Flight
Page 29: © Universal History Archive/UIG via Getty Images
Page 30: Courtesy Jim Ross/NASA
Page 31: Courtesy Ken Ulbrich/NASA; © Evening Standard/Getty Images
Page 32: Courtesy NASA/NASA.gov/Quentin Schwinn: © Rex Features via AP Images
Page 33: Courtesy USAF/NASA
Page 34–35: © USAF/The LIFE Picture Collection/Getty Images

Chapter 3: Higher
Pages 36–37: Courtesy NASA via NASA Johnson Flickr
Pages 38–39: © Doug Hallinen/NASA/Alamy
Page 39: Courtesy Everett Collection; Courtesy K. E. Tsiolkovsky Museum, Kaluga, Russia/NASA; © Sovfoto/UIG via Getty Images
Pages 40–41: Courtesy NASA
Pages 42–43: © Alan Chan — Space Elevator Visualization Group
Page 42: Courtesy Library of Congress; © Science Photo Library/Alamy Stock Photo
Page 44: Courtesy Pat Rawling/NASA
Page 45 (both): © Alan Chan/www.mahalobay.com

Chapter 4: Farther
Pages 50–51: © Dinodia Photos/Alamy
Pages 52–53: © Universal Images Group Limited/Alamy

Page 53: Courtesy Professor Dava Newman, MIT, Inventor, Science and Engineering; Guillermo Trotti, AIA, Trotti and Associates, Inc. (Cambridge, MA), Design; Dougas Sonders, Photography
Page 54: Courtesy NASA
Page 55: Courtesy NASA; © NordicImages/Alamy Stock Photo; Courtesy Leik Myrabo/LightCraft Technologies
Pages 56–57: © National Geographic Image Collection/Alamy
Page 56: © World History Archive/Alamy
Page 57: Courtesy NASA/JPL/Malin Space Science Systems

Chapter 5: Smarter
Page 60–61: Courtesy ESA/NASA
Page 62: Courtesy Planet Labs
Page 63: Courtesy Vicky Somma/Flickr
Page 64–65: Courtesy NASA
Page 64: Courtesy NASA
Page 65: Courtesy Zac Manchester and Ben Bishop/KickSat; Courtesy NASA
Page 66–67: Courtesy ESA; Courtesy NASA
Page 67: Courtesy ESA/Orianne Arnould
Pages 68–69: Courtesy Robert Wood, Harvard University
Pages 70–71: Courtesy NASA

Chapter 6: Dreams Aloft
Pages 72–73: © detchana wangkheeree/Shutterstock
Pages 74–75: © NASA Photo/Alamy
All images pages 76–79: Courtesy Curiosity Machine
Page 80: © Sergey Laurentev/Shutterstock; © bierirderlondon/Shutterstock
Page 81: © Ondrek Prosicky/Shutterstock; © StudioSmart/Shutterstock
Page 83: Courtesy NASA/Pat Rawlings, SAIC
Page 84–85: Courtesy NASA/ESA, SSC/CXC/STScI
Page 86: Courtesy NASA

A cosmic selfie!